Real Science-4-Kids

Teacher's Manual

Level I

Dr. R. W. Keller

PHYSICS

RealScience
4
Kids

Cover design: David Keller
Opening page: David Keller, Rebecca Keller
Illustrations: Rebecca Keller

Copyright © 2004, 2007 Gravitas Publications, Inc.

Real Science-4-Kids: Teacher's Manual

ISBN 10: 0-9749149-7-5
ISBN 13: 9780974914978

Published by Gravitas Publications, Inc.
P.O. Box 4790
Albuquerque, NM 87196-4790

Printed in United States

GRAVITAS
PUBLICATIONS INC

A note from the author

This curriculum is designed to give students both solid science information and hands-on experimentation. This level is geared towards fourth to fifth grades and much of the information in the text is very different from what is taught at this grade level in other textbooks. However, I feel that students beginning in the fourth grade can grasp most of the concepts presented here. This is a *real* science text and so scientific terms are used throughout. It is not important at this time for the students to master the terminology, but it is important that they be exposed to the real terms used to describe science.

Each chapter has two parts: a reading part and an experimental part. In the teacher's text, an estimate is given for the time needed to complete each chapter. It is not important that both the reading portion and the experimental portion be concluded in a single sitting. It may be better to split these into two separate days depending on the interest level of the child, and the energy level of the teacher. Also, questions not addressed in the teacher's manual may arise, and extra time may be required to investigate these questions before proceeding with the experimental section.

Each experiment is a *real* science experiment and not just a demonstration. These are designed to engage the students in an actual scientific investigation. The experiments are simple, but are written the way real scientists actually perform experiments in the laboratory. With this foundation, it is my hope that the students will eventually begin to think of their own experiments and test their own ideas scientifically.

Enjoy!

R. W. Keller

How to use this manual

This teacher's manual provides additional information and answers to the laboratory experiments and review pages for Physics. The additional information is provided as supplementary material in case questions arise while the students are reading the text. It is not necessary for the students to learn this additional material as most of it is beyond the scope of this level. However, the teacher may find it useful for answering questions.

The laboratory section (or Experiments) together with the Review are found at the end of each chapter. All of the experiments have been tested, but it is not unusual for an experiment to fail. Usually repeating an experiment helps both student and teacher see where an error may have been made. However, not all repeated experiments work either. Do not worry if an experiment fails. Encourage the student to troubleshoot and investigate possible errors.

Getting started

The easiest way to follow this curriculum is to have all of the materials needed for each lesson ready before you begin. A small shelf or cupboard or even a plastic bin can be dedicated to holding most of the necessary chemicals and equipment. Those items that need to be fresh are indicated at the beginning of each lesson. The following is a partial list of equipment required for the experiments:

timer or stop watch
wax paper
rubber bands
paper clips
ruler

balance or food scale
pennies
tacks or straight pins
marbles of different sizes
cardboard tube [such as a wrapping paper tube]
pennies
aluminum foil
plastic-coated copper wire
duct tape [or other strong tape]
plastic or rubber rod
balloons
silk fabric
small magnets
iron filings
batteries [1.5 Volts, any size]
electrical wire
small light bulb [light bulb/socket kit (The Wild Goose #WG 16020) works well]
electrical tape
metal rod [such as a long screwdriver]
glass or plastic prism
flashlight
metal can open at both ends
laser pointer
long wooden craft stick

Laboratory Safety

Most of these experiments use household items. However, some items such as iodine are extremely poisonous. Extra care should be taken while working with all chemicals in this series of experiments. Outlined below are some general laboratory precautions that should be applied to the home laboratory:

Never put things in your mouth except if the experiment tells you to. This means that food items should not be eaten unless it is part of the experiment.

Use safety glasses while using glass objects or strong chemicals such as bleach

Wash hands after handling all chemicals.

Use adult supervision while working with iodine, glassware, and any step requiring a stove.

Contents

Materials at a glance

Experiment 1	Experiment 2	Experiment 3	Experiment 4	Experiment 5	Experiment 6	Experiment 7	Experiment 8	Experiment 9	Experiment 10
tennis ball yarn or string (10ft) paper clip marble	slinky paper clips apple lemon or lime banana ruler balance or food scale	stiff cardboard wooden board or plank (3 feet) straight pin or tack scale or balance small to medium-sized toy car banana slices (10-20) pennies	small glass marbles of different size cardboard tube scissors black marking pen ruler two stop watches letter scale or balance	(10-20) copper pennies paper towels aluminum foil salt water voltmeter plastic-coated copper wire, 4"-6" long duct tape (or other strong tape)	small glass jar aluminum foil paper clip duct tape (or other strong tape) plastic or rubber rod balloon silk fabric (2) small magnets iron filings	insulated electrical wire 12 V battery insulating material; foam, plastic, cloth small light bulb electrical tape several small resistors	metal rod electrical wire (10-20) paper clips 12 V battery electrical tape	two prisms flashlight metal can open at both ends aluminum foil rubber band laser pointer long wooden craft stick colored pencils duct tape (or other strong tape)	student selected materials

Chapter 1: What is Physics?

Time Required:

   Text reading - 30 minutes
   Experimental - 1 hour

Experimental Pre-setup:

   NONE

Additional Materials:

   NONE

Overall Objectives:

This chapter will introduce the students to a fundamental concept in physics called "physical laws." The students will also examine the scientific method.

1.1 Introduction

This section introduces observations about the physical world. Ask the students to describe several observations they may have noticed.

For example:

What happens when they pull on the brakes while riding a bicycle? Do the tires stop immediately? Do they skid?

What happens when they throw a ball into the air? Does it reach the clouds? Does it come down in the same spot?

What happens when they turn on a flashlight? How far can they see the light? Can they see the light of a flashlight in the daytime?

Try to get the students to discuss as many observations as they can. There are no "right" answers and, at this point, the reasons why something happens are not important.

1.2 The basic laws of physics

Ask the students what a "law" is, such as a law against driving too fast or a law against stealing. Ask them if these laws are ever broken and if so, why do they get broken.

Ask the students some questions about what they have consistently observed. For example:

Have they ever thrown a ball and it has not come down (except when it gets stuck in a tree)?

Does ice always float?

Does the sun always come up in the morning?

Explain to the students that a law in physics differs from the kinds of laws that govern our country. In physics a law is an overall principle or relationship that remains the same and is not broken.

1.3 How do we get laws?

Discuss with the students how we make laws for our country, city, or state. Discuss with them how the making of city, state, or federal laws involves a long process where several people decide what kinds of laws to make. Explain that, because there are different people making the laws, some laws are different from city to city or state to state. For example, the speed limit is different in different states because not everyone in every state agrees on what the speed limit should be. Explain to the students that governmental laws are laws we make ourselves and, because of this, the law sometimes differs.

Ask the students if they think physical laws are laws we make ourselves. Do physical laws differ from state to state? Do they think that a baseball hit from a ballpark might be able to reach the clouds in Alaska or Hawaii? Will ice float in Arizona, but sink in New Jersey? The answer is "no," a ball will not reach the clouds in Alaska and "yes," ice still floats in New Jersey.

Explain to the students that physical laws are not laws we make up ourselves. They are regularities in the way things behave that scientists have discovered. Explain that the physical world is ordered, reliable and consistent. This orderliness means there are underlying physical laws, or general principles, that we can discover to better understand the world.

Explain to the students that physical laws are described by mathematics. Because the universe is ordered, mathematics can be used to describe precisely the laws that govern it.

1.4 The scientific method

Although scientific investigation began with Aristotle, the foundations for the scientific method were not established until the 13th century with Roger Bacon and further elucidated in the 17th century by Rene Descartes.

The scientific method has 5 steps:

    1) observation
    2) formulating a hypothesis
    3) experimentation
    4) collecting results
    5) drawing conclusions

The first step in the scientific method is observation. Discuss with the students how observations are made. Give some examples of observations using sight, hearing, taste, smell, or touch. For example:

    Salt is poured on icy roads when it snows.
    Lemons are sour. Oranges are sweet.
    The sky is blue.
    Ice cubes float in soda, water, and milk.

    Thunder is loud when lightning is close.
    Steel balls or marbles are sometimes cold but cotton balls are not.

Have the students turn these observations into questions:

    Why is salt poured on icy roads when it snows?
    Why are lemons sour but oranges sweet?
    Why is the sky blue?
    Why do ice cubes float in soda, water, and milk?
    Do ice cubes float in oil?
    Why is thunder loudest when lightning is closest?
    Why are steel balls or marbles sometimes cold, but cotton balls are not?

Explain to the students that, after making observations and asking questions about these observations, the next step in the scientific method is formulating a hypothesis. Hypotheses are guesses. That is, from an observation and the questions about observations, a statement can be made about why something is a certain way. Although a scientist should attempt to make a good guess, hypotheses do not have to be correct. For example:

1) Salt is put on roads to make rubber tires sticky. (Salt is actually used to lower the freezing temperature of ice causing the ice to melt.)

2) Lemons are sour because they have no sugar. (Lemons have some sugar, just less sugar than oranges.)

3) Oranges are sweet because they have sugar.

4) The sky is blue because all of the other colors get absorbed by water in the atmosphere (the sky is actually blue because of light scattering-- called Raleigh scattering).)

5) Ice cubes float because they repel soda, water, and milk. (Ice cubes float because ice is less dense than liquid water.)

6) Ice cubes will not float in oil.

7) Ice cubes will float in oil.

8) Thunder is louder when it is closer because the sound hits our ears sooner and has less chance to go someplace else if we are close. (Thunder travels as a sound wave and   the further we are from the thunder the more the sound gets dampened as it hits molecules in the air while it is traveling.)

9) Steel balls and marbles sometimes feel cold in our hands because they allow heat to exchange and cotton balls do not allow heat to exchange.

These are some examples of hypotheses. Explain to the students that not every hypothesis is correct. Explain to the students that when scientists formulate a hypotheses, they do not already know the correct answers. A scientist is making an educated guess. The next step in the scientific method, designing an experiment, will help determine whether or not the hypothesis is correct.

Explain to the students that because a hypothesis is a guess, a scientist must do something, like design an experiment, to test whether or not the hypothesis is correct.

Discuss with the students how the boy in the text tests for whether salt makes rubber sticky. Or help the students think of ways they might test for sugar in a lemon or test for another hypothesis.   For example:

1) Lemon juice can be collected and the water evaporated. Sugar might be visible in the remaining residue.

2) Lemon juice residue could be compared to orange juice residue.

3) Ice cubes could be placed in several liquids such as water, milk, soda, and oil to determine if ice cubes always float.

Tell the students that sometimes it is difficult to design experiments and not every question can be answered. For example, it may be difficult for a student to design an experiment to test for certain whether or not lemons have sugar. The residue may contain other chemicals that do not allow one to say with certainty whether or not a lemon has sugar. Also explain to the students that there is not necessarily one right way to do an experiment although there are usually certain parts to an experiment that make it better. For example, controls are used to make sure the experimental setup is working properly. Controls can be either positive or negative.

A positive control tells the scientist how the results for an experiment might look if the hypothesis is true. For example, if lemon juice has sugar and if it is possible to evaporate the water from the lemon juice, then a scientist might want to know what sugar evaporated from water might look like. To find out, a positive control of just sugar and water can be used. The scientist would mix sugar and water together, let it evaporate, and then examine the residue. He could then compare this control with his experiment and find out if they look similar. If they do, he might conclude that there is sugar in the lemon juice.

A negative control tells the scientist how the results should not look or if a positive control might be confusing. For example, it may not be easy to tell the difference between salt water and sugar water. A scientist might set up a negative control where salt is used instead of sugar. If evaporated salt water looks similar to evaporated sugar water, then it won't be easy to tell if the residue from the lemon juice is sugar or salt. Some other test is needed like tasting the residue.

Once an experiment has been designed, results are collected as the experiment is carried out. This is the next step in the scientific method. Explain to the students that it is very important that all of the results be recorded. This includes results that the students did not expect. Sometimes major scientific discoveries are found by results that were not at all expected. A good scientist has a keen sense of observation and does not let what he expects to happen determine what he records. Some possible results for lemon juice might be:

1) No residue was found after the water dried.

2) Residue was found.

3) The residue did not taste like anything.

4) The residue tasted salty (or sweet, or sour).

The final step in the scientific method is making a conclusion. In a conclusion the scientist evaluates the results of the experiment and tries to make a statement regarding the hypothesis. For example, if residue was found in lemon juice and the residue tasted sweet, the scientist can conclude that the hypothesis that "lemons have no sugar" may not be correct. At this point, the scientist needs to determine how conclusive the data are and check the reliability of the experimental set up. It is important that the conclusions be valid and not state something that the data haven't shown.

1.5 Summary

Go over the summary statements with the students. Discuss with them any questions they might have.

Experiment 1:   **It's the Law!**   **Date:** _____

Objective: In this experiment we will use the scientific method to determine Newton's First Law of Motion

Materials:

    tennis ball
    yarn or string (10 ft)
    paper clip
    marble

Experiment:

  Part I

1. Take the tennis ball outside, and throw it as far as you can. Observe how the ball travels through the air. Record what you see in the space below.

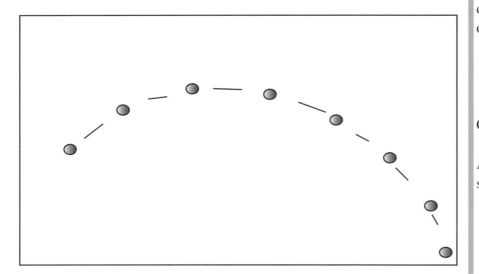

In this experiment the students with discover Newton's First Law of Motion by observing the motion of a tennis ball and a marble.

Newton's First Law of Motion is also called the Law of Inertia. The students will look more carefully at motion and inertia in Chapter 4. However, in this experiment the objective is to show the students that, by observation, they can discover physical laws.

Newton's First Law of Motion can be stated as:

a body will remain at rest or in motion until it is acted on by an outside force.

In the first part of this experiment, the students are to observe how a ball flies through the air. They should notice that the ball will go up and come down in some kind of arc every time they throw it. The arc can be shallow or sharp depending on how they throw it.

Challenge them to throw the ball so that it won't come down.

Ask them if they can get the ball to go up an down in a different pattern such as:

2. Now, take the string or yarn and attach it to the tennis ball using the paper clip. To do this, open the paper clip up on one side and slightly curve the end:

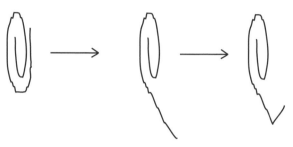

3. Put the extended curved end of the paper clip into the tennis ball by gently pushing and twisting.

4. Next, tie the string to the end of the paper clip.

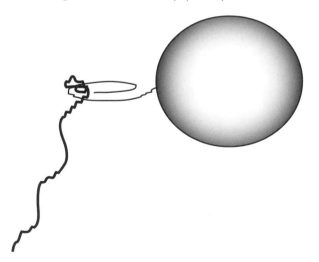

By attaching a string to one end of the ball the students will be able to observe a difference in how the ball will travel once it is thrown.

This is only one way to attach a string to a tennis ball. Several other methods were tried, but it seems that the paper clip works the best. It is somewhat difficult to puncture the tennis ball with the paperclip, so supervise the students. It might help to put a small hole in the tennis ball with a penknife or ice-pick before inserting the paper clip.

If you do not want to puncture the tennis ball, the string can be wrapped several times around the ball and secured with tape.

5. Holding onto one end of the string, throw the ball again into the air as far as you can. Note how the ball travels, and record what you see in the space below. Do this several times.

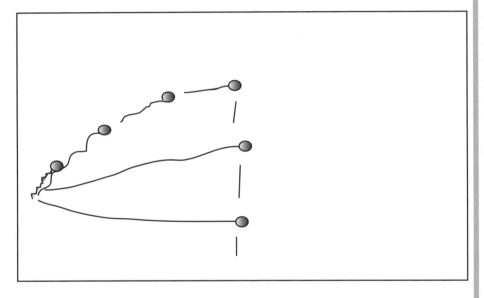

Part II

1. Take the marble and find a straight clear path on the floor or outdoors. Roll the marble on the floor and record how it travels. Note where and how it stops or changes direction. Do this several times and record your observations below.

The trajectory of the tennis ball will now be different. When the students throw the ball, it will begin similarly but when the string has reached its full length, the ball will abruptly stop and fall to the ground.

Have the students throw the ball several times. Ask them if they can change how the ball falls to the ground. They should noticed that, if they shorten the string, the ball does not travel as far as when the string is longer. They should also notice that, if they do not throw the ball very far and the string does not reach its end, the ball will travel almost as if there is no string attached.

Without telling them the answer, help the students see that the ball's trajectory only changes when the string can act on it.

In Part II, the students will examine how a marble rolls. Have them roll the marble on a smooth surface. They should notice the ball traveling mostly straight. Have them compare this with a ball rolling on a rough surface. Discuss with the students why they think the ball may travel differently.

You can also have the students place obstacles in front of the marble, such as small building blocks. They should be able to observe the marble traveling straight on a smooth surface until it contacts an obstacle.

Again, help the students see that the marble's trajectory is not changed unless it is contacted by something -- like a rough surface or a building block.

smooth surface

rough surface

Conclusions:

_____
_____
_____
_____
_____
_____
_____
_____

Help the students make conclusions based on the data they have collected.

Some possible conclusions are:

1) The tennis ball goes up and always comes down.

2) The string keeps the tennis ball from going all the way up because it pulls the tennis ball back.

3) The marble travels on the smooth surface in a straight line but the rough surface keeps the marble from traveling straight.

4) The marble changes direction only when it hits a block.

After the students have thought about their data and drawn some conclusions discuss with them Newton's First Law of Motion. Show them how they were able to observe the same things that Newton observed and that they too could *discover* a fundamental law of physics.

Review

Define the following:

physics   *the study of how things move and behave in nature*

law   *a precise statement about how things behave*

List the 5 steps of the scientific method:

observation

forming a hypothesis

experimentation

collecting results

drawing conclusions

NOTES:

Chapter 2: Force, Energy, and Work

Time Required:

Text reading - 30 minutes
Experimental - 1 hour

Experimental Pre-setup:

NONE

Additional Materials:

Fresh fruit such as a banana, apple, or orange

Overall Objectives:

This chapter will introduce the students to the fundamental concepts of force, energy, and work. These concepts can be difficult to understand. It is not important that the students completely grasp everything about these concepts. This chapter is only a qualitative introduction and the students should be encouraged to think about them but not necessarily understand all of their subtleties.

2.1 Introduction

This section introduces the terms energy, work, and force.

Discuss with the students their own ideas about the terms work, energy, and force.

Ask them:

> What is energy?
> Can energy be created or destroyed?
> What happens to the energy in a battery when the battery dies?
> What is work?
> When you move bricks from the front yard to the back yard- is that work?
> Is lifting a book work?
> Is dropping a book work?
> What is force?
> Can you give some examples of force?

Most of their answers may not be correct and they may have some misconceptions about energy, work, and force. For example, the energy in a battery is not destroyed when it dies. The usable energy has only been converted into a form that cannot be used any longer. This is a common misconception; students will better understand the nature of energy by examining this misconception.

2.2 Force

This section examines the nature of force in more detail. By definition a force is something that:

> changes the position, shape, or speed of an object.

Using this definition, discuss with the students some of the forces they experience every day. For example:

> What happens when they jump?
> What happens when they collide with another player on the football field?
> What happens when they lift a heavy object?
> What happens when they drop a heavy object?
> What happens when they pull or push on a door?
> What happens to a marshmallow when they squeeze it?
> What happens to a steel ball when they squeeze it?

Explain to them that, in all of these examples, forces are acting.

Discuss gravitational force. Gravitational force is an attraction between any two bodies with *mass*.

Mass has two very important roles in physics. First, as in the paragraph above, any two bodies with mass will attract each other by gravitation. So, the amount of mass determines how strong gravitational forces are. Big objects with large masses attract each other much more than smaller objects with small masses. The *weight* of any object is the gravitational attraction between the object and the earh (as measured at the earth's surface). So *weight* is a force, not a mass. Note that the English unit of weight, pounds, is a unit of force while the metric unit, kilograms, is a unit of mass.)

Second, the mass of an object controls its inertia, the tendency of any body to resist movement when acted on by a force. A body with a large mass will accelerate slowly when acted on by a force, but a body with a small mass will accelerate quickly when acted on by the same force. At first glance, this seems like a completely different property than gravitational attraction and it is indeed a very strange fact that one property of matter, mass, controls two seemingly different things. The fact that *gravitational mass* and *inertial mass* are the same is called the Equivalence Principle. The equivalence principle is the basis for Einstein's theory of gravitation (also called General Relativity). Both inertia and gravitation arise from one thing -- the curvature of space itself!

The students will look at inertial mass in more detail in Chapter 4.

The force due to gravity is:

$$F = G * m_1 m_2 / d^2$$

where G is the universal gravitational constant, m_1 is the mass of the first body, m_2 is the mass of the second body, and d is the distance between the two bodies.

We can see from the equation that a body that has more mass will have a greater force. Explain to the students that because their mass is so much less than the mass of the earth their gravitational force is much smaller and so they cannot pull the earth towards them. This is why they come back to the earth when they jump rather than the earth shifting upwards to meet them. (Do they think it might be possible to shift the earth by having everyone jump at the same time? Why or why not?)

2.3 Balanced forces

Explain to the students that, when objects aren't moving, forces are balanced.

The most fundamental equation defining force is Newton's Third Law of Motion:

$$F = ma \text{ or } a = F/m$$

where F is force, m is mass and a is acceleration. This equation says that the force exhibited by an object is equal to its mass times its acceleration. This equation shows that if the acceleration of an object is zero, then the net force acting on that object is also zero. Or conversely, a force acting on an object causes it to accelerate. For a given force, an object of small mass will acclerate more than an object of large mass.

Explain to the students that, in the drawing, a ball is sitting on a shelf. You can see from the diagram that the arrows point in opposite directions. The ball is pushing down on the shelf but at the same time the shelf is pushing up on the ball. Point out to the students that the forces are equal but acting in opposite directions.

In this case, the ball does not move and so the net force is zero.

F_{net} = [force of the ball pushing down] - [force of the shelf pushing up] = zero.

The forces are balanced, i.e., they cancel each other out.

Explain to the students that objects that are moving but not *accelerating* also have balanced forces. Discuss the diagram with the hockey puck and show them that, although the hockey puck is moving, it is not *accelerating*.

2.4 Unbalanced forces

When forces become unbalanced the object will accelerate. Acceleration is the change of an object's speed over time, like a car speeding up from a stop light or a ball speeding up when it is dropped. An object's speed also changes when it slows down, so slowing down is also a form of acceleration. Just as a massive object is hard to speed up, it is also hard to slow down.

Discuss with the students some examples of unbalanced forces:

Does a ball thrown in the air have balanced or unbalanced forces? Why? *unbalanced because it speeds up and slows down.*

Does an airplane when it takes off have balanced or unbalanced forces? *unbalanced - speeds up to take off*

Does a car going from a stoplight have balanced or unbalanced forces? *unbalanced*

2.5 Work

The concept of work may be difficult to understand because when we hear the word "work," we think of mowing the lawn or doing the laundry. However, in physics, work is defined as:

$$work = distance \times force$$

The illustration in the student text shows that, for the same amount of force, the work a short weight lifter does is less than the work a tall weight lifter does because the distance is less for the short weight lifter.

Discuss with the students other examples relating work, distance, and force. For example:

If you carry a box of books up one flight of stairs and your brother carries the same box up two flights of stairs, who has done more work? (*your brother*) How much more work has he done? (*exactly twice the amount of work*)

If you carry a box of books up one flight of stairs and your brother carries a box of books that has half the mass up one flight of stairs, who has done more work? (*you have*) How much more? (*exactly twice as much*)

Help the students think of some of their own examples. The more examples they can discuss the more they will understand the relationship between work, distance, and force.

2.6 Energy

Energy is another concept that can be difficult to grasp because when we hear the word "energy" many different ideas come to mind. Energy basically gives objects the ability to do work. The different kinds of energy, such as potential energy, kinetic energy, and heat energy, will be discussed in more detail in later chapters. What is important for the students at this point, is for them to begin thinking about where the ability for an object to do work comes from.

Ask the students to list some forms of energy that they are familiar with and discuss with them what kind of energy they think might be used.

What do you need to make a car run? *gasoline (chemical energy)*
What do you need for a flashlight? *batteries (chemical energy)*
What do you need for a CD player? plug into an outlet *(electrical energy)*
What do you need to carry books up a flight of stairs? *muscles (mechanical energy), food (chemical energy)*

2.7 Summary

Go over the summary statements with the students. Discuss with them any questions they might have.

NOTES:

Experiment 2:   Fruit Works?   Date: ———————

Objective: _____

_____

Hypothesis: _____

—————————————————————————————————

Materials:

    slinky
    paper clips (2)
    apple
    lemon or lime
    banana
    ruler
    balance or food scale

Experiment:

1. Try to decide, just by "weighing" each piece of fruit in your hands which piece will do the most work and which piece will do the least work on the spring.

2. Write down your prediction stated as the Hypothesis.

3. Now weigh each piece of fruit on the balance or food scale.

4. Record the weights on the chart below.

In this experiment the students will try to determine how much work a variety of fruit can do. Remind the students that:

work = distance x force

Have the students read the entire experiment and then help them think of possible objectives. For example:

> Using a slinky, we will find out if a banana can do more work than an orange.

> We will measure the work fruit can do.

> We will find out if two bananas do more work than one.

Have the students make a guess about which fruit can do more work. They should be able to tell just by weighing the fruit in their hands which one is the heaviest. Have them state these in the Hypothesis. For example:

> A banana is heavier than a lemon and will do more work.

> An orange is lighter than the apple and will do less work.

> Two bananas will do more work than one banana because two bananas weigh more.

Using a food balance or a small scale, have the students weigh each piece of fruit and record their weights in the chart.

Fruit	Weight (oz. or g.)

5. Next, take the paper clip and stretch one side out to make a small hook.

6. Place the hook in one of the pieces of fruit.

7. Hold the slinky up to the level of your chest and allow 10 to 15 coils to exist below. You will have to hold most of the slinky in your hand.

8. Measure the distance from the floor to the bottom of the slinky with the tape measure. Record your result below.

Distance from floor to slinky

_____

9. Now place the piece of fruit that has the hook in it on the slinky and allow the slinky to be pulled out by the fruit.

10. Measure from the end of the slinky to the floor with the tape measure and record your results below.

11. Repeat with each piece of fruit. Record your results below.

After the students have recorded the weights of the fruit, have them use the paper clips to create hooks for the fruit. We found the paper clips worked fairly well, but the younger kids found tape more effective. The fruit can be fixed to the slinky in any manner.

The students will have to experiment with the slinky and number of coils. We found that having the students hold most of the coils in their hands and allowing only a few to fall below worked fairly well. Also, instead of holding the slinky, it can be attached to the branch of a tree or some other fixed ledge. Just make sure the slinky is free to extend and does not contact any other surface.

Have the students first measure the distance from the end of the slinky to the floor without a piece of fruit on it. The distance should be around 2 or 3 feet. Make sure that once this distance is measured, the number of coils allowed to extend is not altered. If the distance is too short (that is the fruit extends the slinky to the ground) reduce the number of coils used and remeasure the distance to the floor.

Have the students fix the fruit to the last coil in the slinky and allow the coils to extend. Have them measure the distance from the ground to the bottom of the piece of fruit.

Fruit	Distance from floor to slinky	Distance extended

Have the students subtract the distance the slinky extended without any fruit on it from the distance it extended with the fruit on it. This will give the net displacement. Have them record this in the column marked "Distance extended."

11. Subtract the distance you recorded in step 8 from each of the distances you measured and recorded above. This gives you the distance each piece of fruit has extended the slinky.

12. Calculate the work each piece of fruit has done. Record your answers in the chart below

Fruit	Work

Have the students calculate the work each piece of fruit has done. They calculate this by using the equation;

$$work = distance \times force$$

Force is the weight of the fruit.

13. What would happen if you put two pieces of fruit on the slinky? Test your prediction and record your answer below.

(2)Fruit	Distance from floor	Distance extended	Work

Now have them predict what would happen if they placed two bananas or two oranges on the slinky. They should predict that two pieces of fruit will do exactly twice the amount of work. Have them test this prediction by fixing two pieces of fruit to the slinky and measuring the distance the slinky extends. Have them record their results.

14. Make some conclusions about your results and record them below.

Conclusions:

_____

_____

_____

_____

_____

_____

_____

Help the students make valid conclusions about their results. Also help them record any problems they may have encountered. For example:

The banana did more work than the orange.

Two bananas did twice the work of one banana.

Two bananas did not do twice the work of one banana.

The slinky extended too far and we could not measure two piecees of fruit.

The apple and orange weighed the same and did the same amount of work.

Challenge question:

In the Review there is a challenge question. Have the students think about whether this would be possible and then help them do a rough calculation.

The mass of the earth is 5.98×10^{24} kg.

The mass of an average human is 66 kg.

According to the US Census the world population is 6,341,930,833. *http://www.census.gov/cgi-bin/ipc/popclockw*

Total mass of people = # people x mass per person = (6,341,930,833) (66kg) = 418,567,434,978 kg $\approx 4 \times 10^{11}$ kg.

Answer = *NO*- there are not enough people.

Review

Define the following terms:

force   something that changes position, shape, or speed
             of an object

work   work = distance × force

energy   gives objects the ability to do work

Circle the correct answer:

Which object has the greater gravitational force:

a banana or a (bowling ball)
a (car) or a bicycle
the moon or the (earth)
the earth or the (sun)

Answer the following questions:

Is a book sitting on a shelf doing work?   *No*
Is a bowling ball crashing into the pins doing work? *Yes*
How much work is done if you lifted a 3 lb box 2 feet? *6 lb-feet*
How much work is done if you lifted a 2 lb box 3 feet? *6 lb-feet*

List some forms of energy

   potential energy   *chemical energy*   *electrical energy*

Challenge:

Do you think if we could get every person on the earth to jump all at once we
could move the earth? Why or why not? Can you do a rough calculation?

NOTES:

Chapter 3: Potential and Kinetic Energy

Time Required:

    Text reading - 30 minutes
    Experimental - 1 hour

Experimental Pre-setup:

    NONE

Additional Materials:

    Banana

Overall Objectives:

In this chapter the students will be introduced to two different types of energy -- *potential energy* and *kinetic energy*. Potential energy is energy that has the potential to do work and kinetic energy is the energy of motion. The main objective of this chapter is to help the students understand that energy exists in different forms and that it is converted from one form to another. In this chapter students will investigate how potential energy is converted into kinetic energy and vice versa.

3.1 Potential energy

Potential energy is energy that has the potential to do work. Discuss with the students the meaning of the word "potential." Make sure they understand that potential energy is already energy, but that it isn't doing any work. Potential energy has the *potential* to do work. Ask the students to give some examples of potential. For example:

> *A child has the potential to become a famous scientist.*

> *A seed has the potential to become a plant.*

> *A puppy has the potential to become a dog.*

In the text, a book on a table is used as an example of potential energy. Explain to the students that this type of energy is formally called

gravitational potential energy

Gravitational potential energy will be given the abbreviation GPE. GPE is the potential energy of an object that is elevated off the ground. Discuss with the students that the amount of GPE an object has equals the amount of work that was needed to lift the object in the first place. Ask the students the following questions:

> In the Review section for Chapter 2, you calculated the work for lifting a 3lb box 2 feet. How much GPE does this box have?
> *6 foot-lbs*

> In the Review section for Chapter 2, you calculated the work for lifting a 2lb box 3 feet. How much GPE does this box have?
> *6 foot-lbs*

3.2 A note about units

When the students calculated work and the GPE for an elevated box, they multiplied two numbers with different units. Explain to the students that a unit describes the type of quantity being used. Discuss with the students that there are different kind of units. There are units for weight, such as lbs and ounces; a unit for mass, grams; units for length, such as inches, feet, and miles; and units for time, such as seconds, minutes, and hours.

Explain to the students that when they multiplied 3 lb x 2 feet they got a unit of foot-lb. It is an awkward unit from the English system, but it is a unit of energy.

There are two different systems for units in common use. In the United States, most people are still taught the English system of units. Most Americans use the English system for measuring weight, length, and volume. Units in this system are feet, inches, miles, oz, lbs, etc. A preferred system of units for science is the metric system. In the metric system the units are divisible by 10 which makes calculating different quantities easier than with the English system.

Metric units include meters for distance, grams for mass, and liters for volume. Discuss the table on page 18 of the student text and show the students the various equivalencies, i.e. how many inches equal a foot, how many feet equal a mile and so on. Show the students how numbers with metric units are easier to calculate than numbers with the English system. For example:

> How many centimeters are in 5 meters? *(500 -- multiply by 100)*
> How many inches are in 5 yards? *(12 inches/foot, 3 feet/yard =*
> *(12 inches/foot) x (3feet/yard) x 5 yards = 180 inches)*

3.3 Types of potential energy

In section 3.1 the students were introduced to potential energy. Gravitational potential energy is only one kind of potential energy. There are other kinds of potential energy such as chemical potential energy, elastic strain potential energy, and nuclear potential energy, to name a few.

Discuss with the students other kinds of potential energy. For example:

> What kind of potential energy is in cereal? *chemical*

> What kind of potential energy is in a battery? *chemical*

> What kind of potential energy is in a rubber band? *strain or*

mechanical

> What kind of potential energy is in a spring? *strain or mechanical*

3.4 Energy is converted

Discuss with the students how energy gets converted. This is an important fundamental concept that the students should know. Energy is neither created nor destroyed -- only converted to other forms of energy. It is important that the students understand that:

Potential energy is useful only when it gets converted to another form of energy.

Ask the students if they can find any useful ways to use different kinds of potential energy without converting the energy into another form, such as using batteries for tree ornaments or jewelry.

3.5 Kinetic energy

Explain to the students that the potential energy of the book on the table gets converted into kinetic energy when it moves from the table and begins to fall. Discuss the Greek word root for kinetic, *kinetikos*, which means "putting into motion." Explain that kinetic energy is the energy of motion.

Have the students think of some other objects that have kinetic energy. Ask them if these things have kinetic energy:

A car going 45 miles/hour *yes*

A tennis ball in motion *yes*

A basketball sitting still on the floor *no*

A toddler who is not sleeping *yes*

A parent at the end of the day *not always*

Explain to the students that the kinetic energy of an object depends on two things -- the mass of the object and the speed of the object. The formula for kinetic energy is:

$$K.E. = 1/2 \; m \; s^2$$

where m is the mass and s is the speed. Explain to the students that KE is proportional to both the mass of the object and its speed. This means that heavier objects will have more KE at a given speed than lighter objects and slower objects will have less KE at a given mass than faster objects.

Also notice that the KE is proportional to half of the mass and the speed squared. This means that there may be much more kinetic energy in a fast- moving toddler than his slow-moving parent!

3.6 Kinetic energy and work

Recall that work is simply the force of an object multiplied by the distance the object is moved. We know that objects having kinetic energy are moving and that, if they hit another object, they can cause the object they hit to move. There is work done when potential energy is converted into kinetic energy and when kinetic energy is converted into other forms of energy, such as heat and sound. The work done on an object equals the change in kinetic energy of that object.

It is not important for the students to completely grasp this concept, only that, when a moving object contacts another object, energy is converted and work is done.

3.7 Summary

Go over the summary statements with the students. Discuss any questions they might have.

Experiment 3:   Smashed Banana   Date: _____

Objective: _____

_____

_____

Hypothesis: _____

_____

_____

Materials:

    stiff cardboard
    board (over 3 feet)
    straight pin or tack
    small scale or balance
    small to medium-sized toy car (1)
    one banana sliced
    10 - 20 pennies

Experiment:

1. Read through the laboratory instructions and then write an objective and a hypothesis for this experiment.

2. Take a portion of the cardboard and make a backing to put the sliced bananas on. Fix a sliced piece of banana to the cardboard near the bottom.

3. Make a ramp with the board. The end of the ramp should meet the sliced banana. Your set up should look like this:

In this experiment the students will convert the gravitational potential energy of a small toy car into kinetic energy and do "work" on a banana.

Have the students read the entire experiment and write an objective. For example:

> *We will measure how much GPE is needed to smash a banana.*

> *We will show that a heavier toy car needs less height (less GRE) to smash a banana.*

Next, have the students write an hypothesis. For example:

> *The toy car will not be able to smash the banana no matter how high the ramp.*

> *The toy car will smash the banana when the ramp is two or three feet high.*

> *The toy car does not have enough mass to smash the banana.*

> *The toy car needs to have at least 50 pennies to smash the banana.*

Have the students assemble the apparatus. It helps if the board is smooth and reasonably straight. A cardboard tube, such as a wrapping paper tube, also works if cut lengthwise and opened up to make a trough. The cars should have good wheels and roll smoothly and easily to reduce friction.

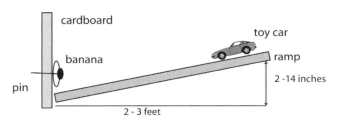

4. Weigh the toy car with the scale or balance. Record your result.

    Weight of toy car (oz. or grams) = _____

5. Place the toy car on the ramp and elevate the ramp 2 inches. Allow the toy car to roll down the ramp and hit the banana. Record your results in the chart below.

6. Elevate the ramp another two inches, now the ramp should be 4 inches off the ground. Allow the toy car to roll down the ramp and hit the banana. Record your results in the chart below

7. Repeat with the ramp elevated 6 inches, 8 inches, 12 inches, and 14 inches and record your results in the chart below.

Distance (inches or centimeters)	Results (write your comments)
2 inches (4 centimeters)	
4 inches (8 centimeters)	
6 inches (12 centimeters)	
8 inches (16 centimeters)	
10 inches (20 centimeters)	
12 inches (24 centimeters)	
14 inches (28 centimeters)	

Answer the following question:

1. At which ramp height did the banana smash? _____

We found that we needed to put several pieces of banana at the bottom because the car often does not travel straight.

Have the students weigh the toy car. Small toy cars typically weigh between one and two ounces.

Have the students roll the toy car down the ramp and then elevate the ramp in 2 inch increments each time allowing the car to travel down the ramp and hit the banana. We found that an average toy car does not really smash the banana until the ramp has been elevated over 12 inches.

8. Now add 10 pennies with tape to the toy car. Weigh the toy car again and repeat rolling the toy car with the pennies down the ramp with the ramp elevated 2 inches, 4 inches, 6 inches, 8 inches, 12 inches, and 14 inches. Record your results.

Weight of toy car plus 10 pennies (oz. or grams) = _____

Distance (inches or centimeters)	Results (write your comments)
2 inches (4 centimeters)	
4 inches (8 centimeters)	
6 inches (12 centimeters)	
8 inches (16 centimeters)	
10 inches (20 centimeters)	
12 inches (24 centimeters)	
14 inches (28 centimeters)	

Answer the following questions:

2. At which ramp height did the banana smash? _____

3. Were the two ramp heights for question 1 and 2 the same? _____

4. If "no" to question 3, which one was higher? _____

Now have the students add pennies to the car to make it heavier. Have the students reweigh the car and repeat the experiment. They should discover that the ramp will not need to be elevated quite as high in order for the toy car to smash the banana.

Using the equation:

gravitational potential energy = weight X height

calculate the gpe for the height that smashes the banana for the toy car with and without the ten pennies. Record your answers below.

gpe for car without pennies _____

gpe for car with pennies   _____

Is the gpe the same [or close to the same] for both cars?_____

Conclusions:

_____

_____

_____

_____

_____

_____

Have the students calculate the GPE for the toy car with and without pennies at the corresponding heights for which the banana was smashed.

The GPEs should be roughly equal. Basically, we expect that it takes a given amount of KE to smash the banana, and it doesn't matter whether this comes in the form of a heavy, slow car or a light, fast car. The energy needed to smash the banana is the GPE the students calculate.

Have the students write conclusions based on the data they have collected.

Review

Define the following terms:

potential energy   *energy that has the potential to do work*

gravitational potential energy *the potential enery of an object*
                                 elevated off the ground

chemical potential energy *potential energy stored in chemicals*

kinetic energy *the energy of moving objects*

Fill in the blanks

1 foot =   *12*   inches

1 yard =   *3*   feet

1 mile =   *1760* yards

1 meter =   *100* centimeters

1 centimeter =   *100* millimeters

1 gram =   *0.001*   kilograms

NOTES:

Chapter 4: Motion

Time Required:

      Text reading - 30 minutes
      Experimental - 1 hour

Experimental Pre-setup:

      NONE

Additional Materials:

      NONE

Overall Objectives:

In this chapter the students will learn about some properties of motion: inertia, friction, and momentum. It is important for the students to understand that an object will remain in a steady, straight line of motion until a force acts on it (review Chapter 2, page 9).

4.1 Motion

Have the students observe or think about objects that move. Ask them why they think objects move. Revisit the experiment in Chapter 1. Ask them what they discovered about objects that move. Have them reread their conclusions. Explain to them that what they observed in the experiment was Newton's First Law of Motion which states:

> *An object in motion will stay in motion unless acted on by*
> *an outside force, and an object at rest will stay at rest unless*
> *acted on by an outside force.*

Discuss with the students the significance of this statement. Explain that for 2000 years people thought that an object that was moving had to have a force pushing it. Explain that Aristotle thought that this was how objects moved, but that he was wrong. Explain to the students that because Aristotle did not feel the movement of the earth, but saw the sun and moon move in the sky, he thought that the earth was the center and that the sun and moon moved around the earth. Discuss the Greek word roots for *geocentric cosmos*.

Have the students look at the drawing of the solar system in the text. Show them where the earth and sun are with respect to each other. Explain to the students that we know today that we live in a *heliocentric cosmos* and that the earth rotates around the sun. Discuss the Greek word roots for this *heliocentric cosmos*.

Explain to the students that there were early scientists who challenged the idea of a geocentric cosmos, but it took 2000 years for people to finally believe that the earth was not the center of the universe. Discuss with the students why they think this might happen. Explain to them that there are a lot of different factors that go into science and that, because science is done by scientists, there are personal biases and pressures that occur. Explain to them that it is sometimes difficult to replace a prevailing scientific theory with a new theory because of these biases and pressures. Explain to the students that this even happens today and new theories are not readily accepted if they are a major challenge to the dominant paradigm.

4.2 Inertia

Discuss with the students the concept of inertia. Explain to them that inertia is the tendency of things to resist a change in motion. Ask them what they think this means.

Ask the students if the following objects are easy or hard to move:

> a toy boat floating on water
> a rowboat on water
> a 20 foot motor boat on water
> an ocean liner on water

Explain to the students that in physics, inertia really means two things: 1) mass and 2) momentum (section 4.5).

The first is the fact that objects with large mass accelerate slowly, when pushed by a force. If you push on a small object with little mass, for example, a toy boat, it accelerates (speeds up) quickly. On the other hand, if you push a larger object, like a good-sized sporting boat, it will eventually begin to move, but you will have to push for a longer time before it accelerates. Explain to the students they could even move a large-sized ocean liner like the Queen Mary by pushing on it and, in the absence of friction, but it would take time. In this case, inertia refers to an object's mass.

4.3 Mass

It is important that the students understand that mass and weight are *different*. Weight is a force. Mass is not. However by weighing an object you can tell how much mass it has since the more mass an object has the more it will weigh on earth because the more force will be exerted on it. Explain to the students that without gravity objects do not weigh anything. In space a boulder floats just like a feather. However, the boulder and feather still have mass. The boulder still has more mass than the feather and, as a result, would still be harder to acclerate than a feather -- even in space.

4.4 Friction

Discuss with the students that although inertia keeps things moving, objects on earth will eventually stop. Ask them what is the only reason anything would stop moving: *because a force acts on the object.* Ask them what force is acting on the following objects to make them stop moving and if they can "see" this force.

      a rolling marble
      a hockey puck on ice
      a car out of gas

Discuss with the students that, although they cannot visibly see why these objects stop moving, a force is still acting on them. Otherwise they would never slow down or stop. For a marble, there is rolling friction between the floor and the marble; for a hockey puck, there is slight friction between the puck and the ice; for a car most of the friction is within the engine, wheels, and axles. (Again, this is what the First Law of Motion states.)

Discuss with the students that this force is friction. Friction occurs when two objects rub against each other. Explain to the students that friction is a force that works in the direction opposite of the direction of motion. Friction is what slows objects down and eventually causes them to stop. Explain to them that, in the absence of friction, an object would keep moving forever and never stop.

4.5 Momentum

The second aspect of inertia is the fact that objects with large momentum are hard to stop.

Review with the students the concept of inertia. Inertia is the tendency of an object to resist a change in motion. Objects that are stationary want to remain stationary and objects that are moving want to stay moving. Review the answers to the questions you asked in section 4.2 -- the two objects on the list that they could stop were:

      a baseball tossed in the air
      a basketball thrown to them.

Now ask them if they could stop the following (without feeling pain):

a baseball hit with a bat

a basketball shot out of a cannon

No is the answer. Now ask them *why*. Have the masses of these two objects changed? *No*. What is different? *The objects are traveling at faster speeds than before.*

Momentum is intertia in motion, that is, mass that is moving. Momentum makes objects hard to stop.

The mathematical equation for momentum is:

$$momentum = mass \times speed$$

Explain to the students that objects that have large masses will have large momentums. Also, objects with fast speeds will have large momentums.

4.6 Summary

Go over the summary statements with the students. Discuss any questions they might have.

NOTES:

Experiment 4:   Moving Marbles   Date:

Objective: _____

_____

Hypothesis: _____

_____

Materials:

several glass marbles of different sizes
steel marbles of different sizes
cardboard tube (2-3 ft. long)
scissors
black marking pen
ruler
letter scale or balance

Experiment:

1. Using the letter scale or balance, weigh each of the marbles, both glass and steel. Label the marbles with numbers or letters or note their colors so that you can know what each marble weighs. Record your results in part "A."

2. Take the cardboard tube and cut it in half lengthwise to make a trough. Mark the middle of the tube with the black marking pen.

3. Measure one foot in either direction of the middle mark and make two more marks, one on each side.

4. The cardboard tube should now have three marks; one in the middle and two on each side one foot from the middle. The tube will be used as a track for the marbles.

Have the students read the entire experiment and write an objective. Possible objectives are:

We will examine the movement of different marbles.

We will investigate the momentum of different marbles.

We will see what happens when one marble hits another.

We will see if we can move a heavy marble with a light one.

We will see if we can move a light marble with a heavy one.

Have them write a hypothesis. Some examples are:

The small glass marble will not be able to move the steel marble.

The small glass marble will be able to move the steel marble.

The small glass marble will stop when it hits the steel marble.

The small glass marble will not stop when it hits the steel marble.

In step 1, the students will weigh the marbles. Remind them that weight and mass are different and that they are not going to know the mass of the objects. However, they will be able to tell which objects *have* more mass -- that is, those that weigh more.

5. Take the marbles and, one by one, roll them down the tube. Notice how each one rolls. Describe how they roll in Results, part "B."

6. Now place a glass marble in the center of the tube.

7. Roll a glass marble of the same size toward the marble in the center. Watch the two marbles as they collide. Record your results in Results, part "C."

8. Repeat steps 6 and 7 with different sized marbles. Record your results in part 'D." [For example, try rolling a heavy marble towards a light marble and a light marble towards a heavy marble].

Have them mark the marbles or note the colors of the marbles so that they will know which marble corresponds to each weight.

Have the students mark the cardboard tube as described in steps 2-4.

In step 5, the students will roll the marbles down the tube and observe how each marble rolls. Have them describe the movement of the marbles in section B of Results. Some examples are:

The glass marbles move easily down the tube and off the end.

The small steel marbles move easily down the tube.

The large steel marble takes more effort to move down the tube.

They may notice that it takes slightly less effort to push the glass marble than the heavy steel marbles. This is because the larger steel marbles have more inertia than the smaller marbles.

In step 6, they will place a glass marble in the center of the tube. When they roll another small glass marble down the tube they should watch closely to see what happens. Have them record their observations in part C. For example:

The rolling glass marble hit the other marble and stopped.

The marble that was stopped started moving when it was hit by the rolling glass marble.

Ask the students to observe the following:

Results:

A.

Marble	Weight

B.

_____

_____

_____

_____

C.

_____

_____

_____

_____

When they roll the glass marble slowly, how far does the other move?

When they roll the glass marble fast, how far does the other move?

They should observe a correlation between how much the second marble moves and how fast the first marble travels. In other words -- fast-moving marbles will cause the stationary marble to move further than slow-moving marbles. Have the students observe this several times.

What they are observing is the *conservation of momentum*. In the absence of angular momentum and friction, the total linear momentum of the traveling marble would be transferred to the stationary marble. However, because the marble is rolling (and so has angular momentum in addition to linear momentum) and because of friction, the second marble does not pick up quite all of the linear momentum of the first. Despite this, the students should be able to observe qualitatively the second marble moving faster when the first marble is traveling faster. This is because the total momentum is conserved - that is, stays the same.

A better way to illustrate this would be to use an air hockey table. If one is available, try the same experiment with two hockey pucks. The advantage of an air table are first that the pucks have little friction, so momentum is better conserved, and second that the pucks are not rolling so there is no angular momentum to complicate things.

Have the students use different-sized marbles. Have them roll a light marble towards a heavy marble and a heavy marble towards a light marble. Help them carefully observe what happens. They should see:

D.

_____

_____

_____

E..

_____

_____

_____

Conclusions

_____

_____

_____

_____

_____

_____

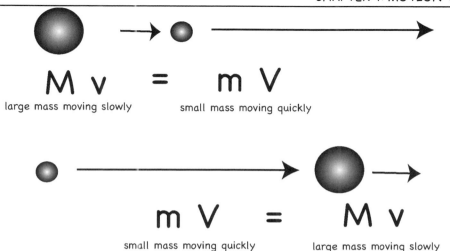

$M\ v\ =\ m\ V$

large mass moving slowly   small mass moving quickly

$m\ V\ =\ M\ v$

small mass moving quickly   large mass moving slowly

When the light marble impacts the heavy marble, the heavy marble will accelerate only a little bit.

When the heavy marble impacts the light marble, the light marble will accelerate quite a bit.

Have the students repeat this several times. Ask them what they think of these results.

Discuss with them the conservation of momentum. Explain that the total momentum stays the same. Remind them that momentum is mass x speed. A large mass traveling at a certain speed will cause a smaller mass to accelerate to a faster speed because momentum is conserved. A small mass traveling at a certain speed will not accelerate a large mass because momentum is conserved.

Although there is angular momentum in a rolling ball and friction present, the conservation of momentum should be observable qualitatively. Have the students draw conclusions based on the data they have collected.

Define the following terms:

inertia *the tendency of things to resist motion*

mass *the property that gives objects inertia*

momentum *the property that makes things hard to stop*

friction *the force experienced by two objects rubbing against each other*

Write the equation for momentum:

momentum = mass x speed

What has more mass; (a bowling ball) or a green pea? (circle one)

What has more momentum: a rolling bowling ball or a (bowling ball) shot from a cannon? (circle one)

NOTES:

Chapter 5: Energy of Atoms and Molecules

Time Required:

    Text reading - 30 minutes
    Experimental - 1 hour

Experimental Pre-setup:

    NONE

Additional Materials:

    NONE

Overall Objectives:

In this chapter the students will learn about the energy of atoms and molecules, that is, *chemical potential energy*, Chemical potential energy is found in fuels, foods, and batteries. Students will also be introduced to nuclear energy. It is important to help the students note that all of the forms of chemical potential energy are useful only when they get converted to other forms of energy like light, heat, or electricity.

5.1 Chemical energy

Discuss chemical energy with the students and have them think about different chemical reactions they may be familiar with. After reading the text, have them discuss what is happening with the jug, vinegar, and baking soda. Explain to them that, as chemical reaction occurs in the jug, chemical energy is being *converted* into mechanical and kinetic energy when the cork pops off the top of the jug. Ask them if they can think of other chemical reactions that get converted into other forms of energy. For example:

> *gasoline in a lawn mower (chemical to mechanical)*
>
> *chemical heating pack (chemical to heat)*
>
> *chemical cooling pack (chemical to heat loss)*
>
> *chemical light stick (chemical to light)*
>
> *matches (chemical to heat)*

5.2 Stored chemical energy

Discuss with the students the fact that chemical energy starts as chemical potential energy before it is converted into other forms of energy. In the text have the students look closely at a steam engine fueled by wood. Explain to the students that, in this illustration, the stored chemical energy in the wood and oxygen gets converted into heat energy when it burns. The heat energy is used to heat water which expands as steam. The work done by the expanding steam is converted into mechanical energy as the piston moves up and down and the train moves forward. In this illustration the energy stored as chemical potential energy is used to convert several different forms of energy from one to another. Have the students look up how cars use gasoline and discuss with them any similarities or differences.

Similarities:

> *gasoline is burned to produce heat and expanding gases*
> *pistons are used*
> *the chemical energy in the gasoline and oxygen is converted into mechanical energy*

Differences:

> *water is not heated*
> *the air in a gasoline engine is compressed*
> *the gasoline is burned inside the cylinders, not outside*

5.3 Stored chemical energy in food

Discuss with the students that chemical energy is stored in food. Have them look back at Chapter 8 Chemistry, Level I and review the *chemical energy* found in certain foods. Discuss with them which foods have carbohydrates. For example:

potatoes
bread
sugar
spaghetti

Explain to the students that the body uses stored chemical energy in foods for energy. The chemical reactions that occur inside the body are complicated but, overall, they amount to essentially the same thing as burning sugar. Explain that many different kinds of reactions take place, but that, overall, the foods we eat get burned and turned into other forms of energy such as heat and mechanical energy (mainly heat). Explain to the students that our bodies require a continual supply of energy foods because unlike plants, we cannot produce our own food. Ask them, "What is the one source of energy on which all plants and animals depend?" *the sun*

5.4 Stored chemical energy in batteries

Another form of stored chemical energy is found in batteries. Explain to the students that a battery is specifically designed to convert chemical energy into electrical energy. Ask the students to list items they are familiar with that use batteries:

CD player
flashlight
many different battery-operated toys
automobiles

The first battery was invented by Alessandro Volta. He constructed a voltaic cell which used alternating layers of metals and salt water to generate electrical energy. The types of batteries used in CD players and flashlights are called dry cells (see Chapter 6).

A simple battery can be made with a lemon. A lemon, some copper wire, and a paper clip are all that are needed to construct a small battery which can light a small light bulb.

5.5 Nuclear energy

Explain to the students that another form of stored energy in atoms is *nuclear energy*. Nuclear energy is released when the nuclei of atoms split into smaller pieces or when smaller nuclei combine into bigger ones. Explain to the students that nuclear reactions differ significantly from chemical reactions. In nuclear reactions the atoms themselves change their identities (say from C to N), but in chemical reactions the atoms only change which other atoms they are bonded to.

Discuss with the students how an atom can change. Explain to them that, by changing the number of protons in an atom's nucleus, the element changes. Discuss the example showing how a nitrogen atom gets converted into a carbon atom by losing a proton. The carbon atom is called carbon-14 (14 = 6 protons + 8 neutrons). Have the students look on the periodic chart (from Chemistry Level I) and ask them how many protons and neutrons a carbon atom has. Show them that a carbon-14 atom still has 6 protons, but has 8 neutrons, two more than normal carbon. Carbon-14 is called an isotope of carbon.

Nuclear reactions release much more energy than chemical reactions. Nuclear energy is used to power nuclear reactors. Explain to the students the design of a nuclear reactor. Show them that the nuclear energy produces heat energy that is then used to heat water. The steam from the water turns a turbine that generates electrical energy.

5.6 Summary

Go over the summary statements with the students. Discuss any questions they might have.

NOTES:

_____
_____
_____
_____
_____
_____
_____
_____
_____
_____
_____
_____
_____
_____
_____
_____
_____
_____
_____
_____
_____
_____
_____
_____
_____
_____
_____
_____
_____

Experiment 5:   Power Pennies   Date:_____

Objective: _____

_____

Hypothesis:_____

_____

Materials:

    10-20 copper pennies
    paper towels
    aluminum foil
    salt water (2-3 T. per cup)
    voltmeter
    plastic-coated copper wire, 4"-6" long
    strong tape (duct tape)

Experiment:

1. Cut out several penny-sized circles from the aluminum foil and paper towel.
2. Soak the paper-towel circles in salt water.
3. Strip off the end of one of the pieces of wire. Tape the exposed metal to a penny.
4. Strip off the end of another piece of wire. Tape the exposed metal to a piece of aluminum foil.
5. Putting the aluminum foil with the wire on the bottom, place one of the wet paper-towel circles on the aluminum foil. Place the penny with the taped wire on top. It should look like this:

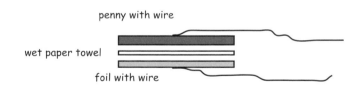

penny with wire

wet paper towel

foil with wire

Have the students read the entire experiment before writing the Objective and Hypothesis.

Some examples are:

Objective:

To discover how a simple voltaic cell operates.

To construct a voltaic cell and measure voltages.

To see if pennies, aluminum, and salt water can really make electricity.

Hypothesis:

Pennies, aluminum foil, and salt water will not generate electricity.

Pennies, aluminum foil, and salt water will generate electricity.

More layers in the voltaic cell will generate more electricity.

Assemble all of the materials before starting. It helps to scrub the pennies with steel wool. Help the students cut out small, penny-sized circles of aluminum foil and paper. It is important that the cutouts be very close to the size of the penny. Soak the paper circles in the salt water.

6. Now take the wires and connect them to the leads of the voltmeter. Switch the voltmeter to "voltage" and record the number. This is the amount of voltage the single layer battery produces.

7. Add another "cell" to the battery and record the voltage. [a cell is a penny layer, a foil layer and a paper layer] It now has two cells. The battery should look like this:

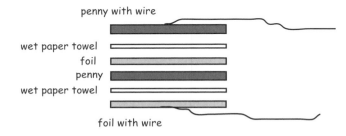

penny with wire

wet paper towel

foil

penny

wet paper towel

foil with wire

8. Continue adding cells of foil, wet paper towel, and pennies and record the voltage for each newly added cell.

Results:

Cell	Voltage
1	
2	
3	
4	
5	

Help the students carefully strip the plastic off of the wire. It can be difficult to strip just the plastic and not cut the wire itself. The best way is to use a pair of wire cutters. Gently squeeze and pull the plastic off the end of the wire.

Have the students tape the exposed wire to one penny and one circle of aluminum foil. These will remain as ends. Additional layers will be added in between these two ends.

Now have the students carefully insert one of the salty paper circles in between the copper penny and aluminum foil. It helps to set one side on a firm surface, add the salty paper towel circle, and then place the other side on top, holding it down with your fingers.

Have the students hook the leads to a voltmeter and read the voltage. An inexpensive voltmeter can be purchased at any store which supplies electrical equipment. Carefully read the instructions for the voltmeter. Make sure that the voltmeter is set to "voltage" and that voltage scale is low enough to detect small voltages. (A typical penny-cell produces about 0.5 V)

Have the students record the voltage they read on the voltmeter. Next have them add additional "cells" to the battery. A cell consists of a penny layer, a soaked paper layer, and an aluminum layer.

one "cell"

Plot your data. Make a graph with voltage on the x-axis and # of cells on the y-axis.

Have the students record the voltage with each new added cell.

The students should continue to add additional cells and record the voltages for as many cells as they can.

Have the students plot their data. The voltage should be on the x-axis and the # of cells on the y-axis.

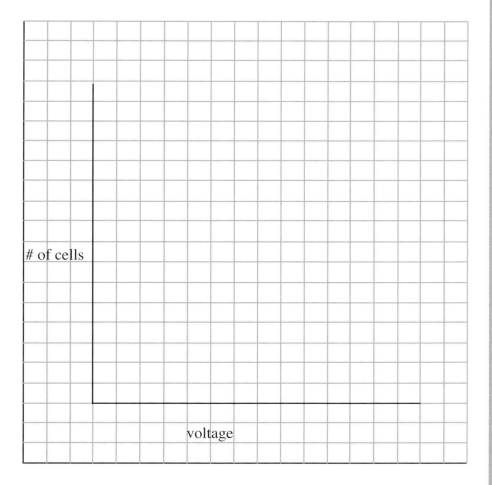

# of cells

voltage

Discuss your data below:

_____

_____

_____

_____

_____

_____

_____

_____

_____

_____

_____

_____

Conclusions

_____

_____

_____

_____

_____

_____

Have the students discuss their data. They should observe that, overall, as the number of cells increases, the voltage increases. However, if there are places where the paper towel or aluminum foil is larger than the penny and overlaps and touches a previous layer, there will be a "short"- that is the electricity will go through the shortest distance. This means that it is possible to short circuit the battery. If part of one cell touches a previous cell, one of the two cells will not be counted. If this happens the voltage will fluctuate. The voltage will appear to increase and then decrease. Have the students discuss these "sources of error."

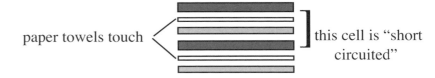

paper towels touch   this cell is "short circuited"

Have the students draw conclusions based on the data they have collected.

NOTE:

It is also possible to use vinegar instead of salt water. If time permits, have the students repeat the experiment using vinegar and compare results with the salt water.

Review

Answer the following questions:

1. What is chemical energy?

　　the energy released from chemical reactions

2. Give two foods that have "food energy" (carbohydrates)

　　　　potatoes

　　　　pasta

3. Give two examples of energy we use for fuel.

　　　　gasoline (wood, coal, sunlight, etc.)

　　　　nuclear reactions

4. Who made the first battery?

　　　　Alessandro Volta

5. Draw a diagram of a voltaic battery:

Penny with wire

wet paper towel

foil with wire

NOTES:

Chapter 6: Electrical Energy and Charge

Time Required:

    Text reading - 30 minutes
    Experimental - 1 hour

Experimental Pre-setup:

    NONE

Additional Materials:

    NONE

Overall Objectives:

In this chapter the students will learn about electrical energy. They will review the structure of an atom and examine in more detail the nature of electric charge.

6.1 Electrical energy

In Chapter 5 the students were introduced to electrical energy that is produced with batteries. In the experiment for Chapter 5, the students built a voltaic battery and discovered that, as they added more cells, more voltage was produced.

Discuss with the students other types of batteries. The type of battery used in most small electronic equipment is called a "dry cell" battery. Have the students look at some dry cell batteries. DO NOT allow the students to break open a dry cell battery. These batteries contain caustic materials that can burn skin.

Have the students look at different sizes and shapes of dry cell batteries and compare the voltages. Ask them why they sometimes need to put more than one battery in a toy or CD player. Ask them what they notice about how the batteries are positioned in a typical electronic toy.

Ask the students if there are other objects that require electrical energy but that do not use batteries. For example:

> desktop computer
> microwave oven
> washing machine
> hair dryer
> power saw
> electric drill

Ask the students if they are aware of other forms of electrical energy (*static electricity, lightning, etc.*).

6.2 Electric charge

Have the students review the parts of an atom. Review Chemistry Level I, Chapter 2, and explain that the parts of an atom that "carry" electric charges are the electrons and protons. Electrons and protons carry charges, but neutrons have no charge. The "sign" of the charge for protons and electrons (positive or negative) is largely arbitrarily assigned. That is, by convention, electrons are negative and protons are positive. The important aspect is that they are *oppositely* charged.

Explain to the students that

> *opposite charges attract*
> and
> *like charges repel.*

Have them look closely at the diagram of the atom. Show them that, for a helium atom, there are two electrons and two protons. Explain to the students that the two electrons in a helium atom will try to be as far away as possible from each other. Explain to them that the positive charges of the protons keeps the electrons from flying away from the atom.

6.3 Charging objects

Have the students list some times when they have observed objects being charged. For example:

pulling clothes out of the dryer

rubbing a balloon and sticking it to a wall

dragging their stocking feet on the carpet

Explain to the students that, in any of these cases, objects are being charged by electrons being moved from one object to another. Explain to them that the fact that the objects are charged becomes visible when they observe two objects "sticking" to each other (such as socks coming from a dryer sticking to the bedsheets), or when objects repel each other (such as the individual hairs on their head separating themselves from each other and sticking outward).

6.4 Electric force

Discuss electric force with the students. Explain to them that electric force is like other forces because electric force causes a change in shape or speed of something such as the movement of charged particles like electrons.

Ask the students to list the ways electric forces change the position of an object: Some examples are:

A charged balloon will pull objects, such as oppositely charged hair, towards it.

A charged bedsheet will drag a sock out of the dryer.

Plastic peanuts become irritatingly charged and are difficult to remove.

Ask the students if they think they could make a simple instrument that could detect electric charge. In the experimental section, the students will construct a simple electroscope where they will detect electric charge.

6.5 Summary

Discuss with the students the summary statements.

Experiment 6:   Charge It!   Date: _____

Objective:

_____

_____

Hypothesis:

_____

_____

Materials:

      small glass jar
      aluminum foil
      paper clip
      strong tape (duct tape)
      plastic or rubber rod
      balloon
      silk fabric

Experiment:

A. Building an *electroscope* [an instrument that detects electric charge]

   1. Cut two thin strips of aluminum foil of equal length about 1.0" long.

   2. Poke a small hole in the center of the lid of the glass jar.

   3. Open one end of the paper clip to make a small hook.

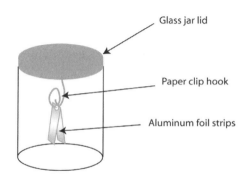

Glass jar lid

Paper clip hook

Aluminum foil strips

Have the students read the experiment and then have them write an Objective and a Hypothesis.

Some examples are:

Objective:

    We will build an instrument to detect electric charge.

    We will test for electric charge with an electroscope.

Hypothesis:

Have the students guess what might happen with the electroscope. Give them the hint that the two pieces of aluminum foil will be the same charge- What will the aluminum foil do?

    The aluminum foil pieces will separate in the electroscope.

    The aluminum foil pieces will stick to each other in the electroscope.

Have the students assemble the parts for the electroscope. Assist the students as they puncture a small hole in the top of the jar.

Have them place the aluminum foil pieces on the paperclip and insert this into the jar.

4. Place the straight piece of the hook through the small hole in the glass jar lid and secure it to the lid with strong tape. Leave the end exposed.

5. Hang the two strips of aluminum foil from the hook and place inside the jar.

6. You now have an electroscope.

7. Take the plastic or rubber rod and rub with the silk fabric, or take the balloon and rub it in your hair.

8. Gently touch the balloon or plastic or rubber rod onto the paper clip that is sticking out from the glass lid.

9. Observe the two pieces of aluminum foil and record your results.

Results:

_____

_____

_____

_____

_____

_____

_____

_____

_____

_____

_____

_____

Once the electroscope is assembled, have the students rub a plastic rod (a plastic comb will work) against a silk cloth or in their hair. They should be able to charge the plastic rod.

Have them gently touch the rod to the tip of the paper clip that is sticking out from the jar.

Have them observe what is happening to the two aluminum foil pieces. They should see them separate. Have them record what they see. Have them think about the following:

What happens if they touch their fingers to the end of the paper clip?

How long does the charge last?

Are there things that make the aluminum foil separate further? More rubbing? Different plastic rod? A glass rod? A metal rod?

How the electroscope works:

Like charges repel, so on a charged object like a balloon (or the parts of an electroscope) the charges always spread out as far apart as they can. When you touch a charged object like the plastic rod to the paper clip in the electroscope, the electrons spread from the rod to the clip and aluminum strips. Now the strips both have negative charges so they repel. After awhile the charge leaks aways, and the strips come back

Conclusions

_____

_____

_____

_____

_____

_____

_____

_____

_____

_____

_____

_____

_____

together. The bigger the charge, the more the strips repel, so you can tell how strongly charged the rod was to start with.

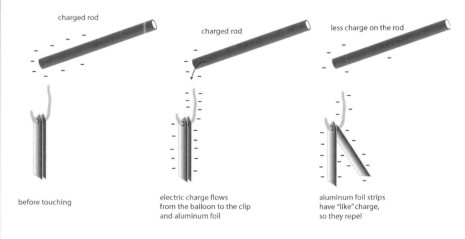

charged rod

charged rod

less charge on the rod

before touching

electric charge flows
from the balloon to the clip
and aluminum foil

aluminum foil strips
have "like" charge,
so they repel

Help the students write conclusions based on the data they have collected.

Review

Define the following terms:

    dry cell   _batteries that use pastes instead of liquids_

    electric charge _the charge assigned to protons and electrons_

    electrical force _force as a result of electric charge_

Circle the correct word to complete the statement:

    Like charges (repel) or attract each other.

    Unlike charges repel or (attract) each other.

List the parts of an atom and whether or not they are charged:

        proton- positively charged

        neutron - no charge

        electron - negatively charged

NOTES:

Chapter 7: Moving Electric Charges and Heat

Time Required:

      Text reading - 30 minutes
      Experimental - 1 hour

Experimental Pre-setup:

      NONE

Additional Materials:

      NONE

Overall Objectives:

In this chapter the students will learn about electric current, resistance, and heat. It is important for the students to understand that charges move, or flow, through wires as a result of "electric pressure." This "pressure" is the result of a difference in potential (or voltage) in an electric circuit.

7.1 Moving electric charges

Explain to the students that in the last chapter they observed the effects of static charges in the electroscope. Static charges can be transferred from object to object, but once they are in place, static charges do not flow.

Ask the students the following questions:

How long does it takes for a light to light up once the light switch is turned on? (*immediately*)

How long does it take for the TV to turn on when you plug it in? (*immediately*)

What if you take an extension cord and make the cord very long. How long does it take to turn on the power drill? (*immediately - even with an extension cord*)

Does the length of the cord matter. (*no*)

Discuss with the students that, when charges move, there is an electric current and that this current behaves much like a hose filled with water. The pressure the water hose experiences when the faucet is turned on causes the water to be pushed out of the hose. In fact, if the hose is completely filled with water, water will immediately flow out the other end. This is similar to how electric current works.

Explain to the students that electric pressure is called voltage. The higher the voltage, the more electric pressure, and the more electrons can be moved.

Explain to the students that wires are made of metals and metals conduct electricity, that is, there are plenty of electrons inside metals that can move. When an electrical pressure is applied to a wire, the electrons inside the wire all move. Explain to the students that this is why the length of the cord does not matter. Electrical pressure at one end pushes electrons out the other end immediately. So the light switch, TV, and power drill all turn on immediately when the cord is plugged into the wall.

Examine with the students the diagram showing several atoms in section 7.1. Show them that the electrons move but the protons and neutrons do not.

7.2 Resistance

Materials that allow electrons to flow easily are called *conductors*. Metals are good conductors because they have lots of electrons that are free to move from atom to atom. Materials that don't allow electrons to move easily from atom to atom are called *insulators*. Insulators are *resistant* to electric flow.

Electrons can move from atom to atom only if there is a place in the receiving atom to accommodate the incoming electron. Insulators do not have space in their atomic shells to accommodate extra electrons, so electrons do not move.

Explain to the students that in most electrical circuits there are small components called resistors. Resistors are used to slow down the flow of electrons, controlling how much electrical current flows.

Typical resistors look like this:

The stripes show the resistance value for the resistor.

7.3 Heat

When lots of electric current flows through a wire, the wire often feels warm or hot to the touch.

Ask the students to describe some things that are heated electrically. For example:

> *burners in a stove top*
> *filament in a light bulb*
> *toaster*
> *electric blanket*

Ask the students where they think the heat comes from in these cases.

Explain to the students that when electrons flow through a metal the electrons collide with the atoms and impart some kinetic energy to the atoms. The atoms shake and vibrate, and we experience this extra energy as heat. The more electrical current that flows thorugh the metal, the more it is heated. Some materials get so hot they glow, like a red-hot burner on a stove top, or the white-hot filament in a light bulb.

Have the students list some objects that do not get hot. For example:

> *styrofoam*
> *certain plastics*
> *wood*
> *cloth*

Explain to the students that these are all insulators - they do not conduct electricity and no electrical current flows through them.

7.4 Summary

Review the summary statements with the students.

Experiment 7:   Let it Flow   Date: _____

Objective: _____

_____

Hypothesis: _____

_____

Materials:
    insulated electrical wire
    several 1.5 V batteries (any size)
    some insulating materials (e.g. foam, plastic, cloth)
    small light bulb
    electrical tape
    several small resistors

Experiment:

1. Cut the wire into two foot-long pieces. Carefully shave off the ends of the plastic insulation to expose the metal. Leave about 1/4 to 1/2 inch of exposed metal on each end.

2. Tape one end of one wire to the (+) terminal of the battery. Tape one end of the other wire to the (-) terminal of the battery.

3. Tape the other ends of the two wires to the light bulb. One wire should be taped to the bottom of the bulb and the other one should be taped to the metal side of the bulb.

4. Record your results.

5. Now place a piece of foam or plastic in between the wire and the bulb.

5. Record your results.

6. Remove one end of the wire from the battery and gently touch the ends of the wire with your finger to see if it is warm.

Have the students read the entire experiment and then write an Objective and a Hypothesis. Some examples are:

Objective:

We will find out if materials such as foam or plastic conduct electricity.

We will observe electric current and find out what happens if an insulating material interrupts the flow.

Hypothesis:

The insulators will not affect the electric flow.

The insulators will keep the light bulb from lighting.

Materials:

optional: Lamp socket and Bulb kit # 16020, The Wild Goose

Have the students carefully strip the plastic off of both ends of the metal wire leaving the ends exposed, then either tape or wrap the metal wire around the two terminals of the battery.

Next, have them tape the other ends of the wires to a small light bulb. One wire should be taped to the bottom of the bulb and the other should be taped to the metal side. It should look like this:

7. Record your results.

8. Now place a resistor in between the light bulb and the battery on one wire. Observe any difference in the intensity of the light bulb. Record your results below. Repeat with two or more resistors.

Results:

	wire only	wire + resistor(s)	wire + insulator
light bulb intensity			
temperature of wire			

Answer the following questions about your experiment:

1. What happened when you connected the battery to the light bulb?

_____

_____

2. What happened when you put a piece of foam or plastic in between the wire and the bulb?

_____

_____

3. What happend when you put one or more resistors in between the light bulb and the battery?

_____

_____

4. What did the wire feel like to your fingers (with the wire only, with the resistors, with the insulator?

_____

_____

Have the students record their results and answer the questions.

They should notice that the light bulb intensity decreases as they add resistors to the circuit. Explain to them that the resistors are behaving like kinks in a hose. They are cutting off the flow of electrons like a kink in a hose cuts off the flow of water. The more resistors added to the circuit, the dimmer the light in the bulb will become.

They should also notice that, as more resistors are added, there is less heat in the wire.

When an insulator is placed between the light bulb and battery, they should find that there is no light coming out of the light bulb.

3. What did the wire feel like to your fingers?

_____

Conclusions:

_____
_____
_____
_____
_____
_____
_____
_____
_____
_____
_____
_____
_____
_____

Have them record their conclusions.

Review

Define the following;

static electricity   *charges that are not moving*

electric current   *charges that are moving*

voltage   *electric pressure*

resistance   *to resist or stand against; a resistor*
                           resists electric flow

conductor   *material that allows electric flow*

insulator   *material that resists electric flow*

heat   *the transfer of heat energy from one object*
                           to another because of a difference in
                           temperature

NOTES:

Chapter 8: Magnets and Electromagnets

Time Required:

> Text reading - 30 minutes
> Experimental - 1 hour

Experimental Pre-setup:

> NONE

Additional Materials:

> NONE

Overall Objectives:

In this chapter the students will be introduced to magnets and their properties. Help the students understand that magnetic fields are produced in two ways: 1) by spinning electrons, and 2) by moving charges, i.e electric currents. The electrons don't hop from one atom to another as in the case of electric current but, instead, spin on their axes. One important concept the students should understand is that electric currents create magnetic fields and magnetic fields can induce electric currents.

8.1 Magnets

During this reading, it would be helpful for the students to have a few magnets to play with.

Ask the students the following questions;

What happens if one magnet is brought close to another magnet? (*the other magnet will either be attracted or repelled*)

What happens if you switch the direction of the first magnet? (*flip it over or use the opposite side?- the magnets will do the opposite from what they did in the first question*)

What happens if you place your magnet near the refrigerator? (*it will stick*)

Does your magnet stick to aluminum cans? (*No*)

Explain to the students that magnets have *poles*. Poles are not "charged" like the balloon or pieces of aluminum foil from the experiment in Chapter 7, but a pair of poles is much like pairs of opposite charges. are opposite fields that produce force.

Explain to the students that in a permanent magnet, poles are caused by moving electrons except, in a magnet, the electrons do not move from atom to atom but, instead, stay in one place and *spin*.

Discuss with the students that electrons spin in a magnet much like a basketball can be spun on their finger. It rotates around an axis and can spin in either direction.

All materials have spinning electrons, but not all materials are magnetic. Why? Explain to the students that electrons in all materials spin in both directions. In materials where the atoms share an even number of electrons the spins "cancel" each other out and the material is not magnetic. However, in materials where there an uneven number of electrons spinning, the extra spinning electrons can make the material magnetic. The extra electrons can align themselves and create *poles*.

Explain to the students that the same "rules" apply to magnetic poles that applied to electric charges:

Like poles repel each other.

Unlike poles attract each other.

It is important for the students to understand that, unlike electric charges,

magnetic poles cannot be separated.

If a thin magnet is available that is easy to cut, it would be useful to have the students try to separate the poles by cutting it again and again until it is very small. Each little piece will still have two poles because the atoms themselves are little magnets.

8.2 Magnetic fields

Have the students slowly bring one magnet close to another magnet. If the opposite poles of each magnet are brought close to each other they should experience a pull as the magnet approaches but before the magnets actually touch. Ask them why they think this might be happening?

Explain to the students that the magnets produce magnetic fields that extend out from the magnet into the space around it. A magnetic field affects the space surrounding the magnet. Even though the magnet physically ends at its edges, the magnetic fields extend beyond these physical ends.

There is an easy way to "see" the magnetic fields and, if time permits, this simple demonstration should be performed.

Take a magnet and some iron filings. An easy way to collect iron filings is to place a magnet in a plastic bag and drag the plastic bag through some dirt. The iron filings will collect on the outside of the bag. Place the bag and magnet inside another bag and remove the magnet from the inner bag. The iron filings will release from the inner bag and collect inside the outer bag. Repeat this several times until a teaspoon or so of iron filings have been collected.

Mix the iron filings with a 1/4 cup of corn syrup and pour the mixture into a clear shallow dish (the bottom portion of a clear glass butter dish works well). Place the dish on top of a regular-sized magnet. Have the students observe the iron filings and then wait for an hour and observe them again. The iron filings should align and look similar to the magnetic field lines shown in section 8.2.

Explain to the students that the magnetic fields pass through the corn syrup without affecting the syrup but, because iron responds to magnetic fields, the iron filings line up with the magnetic field lines.

Explain to the students that magnets can make other objects temporarily magnetic. Using an iron nail, the students can induce a magnetic field in the nail by having the nail contact the magnet. When the magnet is removed, the iron nail will remain magnetic for some time.

8.3 Electromagnets

In the experiment for this chapter the students will build a small electromagnet. Explain to the students that when an electric current flows through a wire it creates a magnetic field around it. When a wire is coiled, it can behave much like a bar magnet. The more electric current that passes through the wire, the stronger the electromagnet.

Explain to the students that electromagnets can be quite strong and are often used in junk yards to lift heavy items such as cars. They are also convenient because they can be "turned" off.

8.4 Electromagnetic induction

Explain to the students that a magnet can also cause an electric current to flow through a coiled wire. In the experiment for this chapter they will see how a coiled wire will create magnetic fields, but a magnet can also induce a voltage in a coiled wire. This is called *electromagnetic induction*. Faraday's Law describes electromagnetic induction. Faraday's Law says:

The induced voltage in a coil is proportional to the number of loops times the rate at which the magnetic field changes within those loops.

This means that the more loops, the greater the voltage, and the faster the magnet is pulled back and forth through the loops, the greater the voltage.

8.5 Summary

Discuss the summary statements with the students.

NOTES:

Experiment 8:   Wrap It Up!   Date: _____

Objective: _____
           _____

Hypothesis: _____
            _____

Materials:
        metal rod
        electrical wire
        paper clips
        12 V battery
        electrical tape

Experiment:

1. Cut the metal wire so that it is one to two feet long.

2. Trim the ends of the wire so that there is 1/4 inch exposed metal.

3. Tape one end of the wire to the (+) terminal of the battery.

4. Tape the other end of the wire to the (-) terminal of the battery.

5. Take the metal rod and touch it to the paper clips. Record your results.

6. Now coil the wire around the metal rod. The wire must remain hooked to the battery.

7. Touch the metal rod to the paper clips. Count the coils and record your results.

8. Wrap another 1 to 5 coils to the metal rod.

Have the students read the experiment and then write an objective and a hypothesis. Some examples are:

Objective:

  In this experiment we will build an electromagnet.

  In this experiment we will explore the properties of electromagnets.

Hypothesis:

  It won't matter how many coils we use, we won't be able to pick up more than a few paper clips.

  The more coils wrapped around the rod the stronger the electromagnet.

  The number of paper clips we pick up will be proportional to the number of coils in the electromagnet.

Have the students assemble the parts for the electromagnet. A screwdrive works well as the metal rod. It should not be magnetized ahead of time. Help the students trim the ends of the wire and fix the ends to the battery. Different-sized batteries can be used, but at big 12 V battery works best. Whenever the wire is connected to both the + and - terminals, the battery is running down, so don't leave it connected when it is not in use.

9. Touch the end of the metal rod to the paper clips. Record how many paper clips can be picked up.

10. Continue adding coils to the metal rod and counting the number of paper clips that can be picked up.

11. Record your results.

Results:

number of coils	number of paper clips

Have the students touch the rod to a pile of paper clips and record the number of paper clips the rod will pick up. Have the students wrap the wire around the metal rod and touch the metal rod to the pile of paper clips.

Have them continue recording the number of paper clips as the number of coils increases.

Using a 12V battery, one student got the following results:

number of coils	number of paper clips
10	very weak
15	1
20	2
25	8
30	12
35	15

Your results may vary, but there should be an overall trend that, as the number of coils increases, the number of paper clips the electromagnet can pick up also increases.

Help the students graph their results.

Graph your results below:

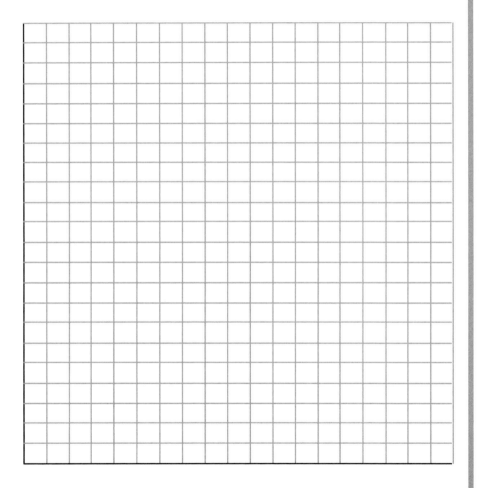

The physics behind an electromagnet:

When a current flows down a wire, a magnetic field is created that rolls around the wire:

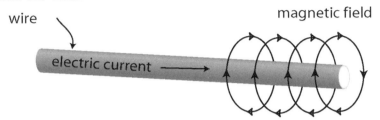

The strength of the magnetic field is proportional to the current; when there is more current, there is a stronger field. If you now wind a wire into a coil, the fields from each part of the coil add up to create a net magnetic field that looks much like the field of a bar magnet:

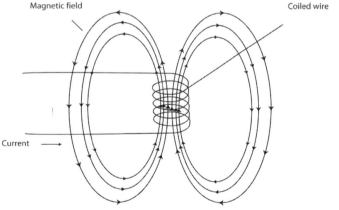

It can be shown that the strength of the magnetic field near the coil is proportional to the number of loops in the coil, more loops -- stronger field.   So the students should see a stronger magnet (one that picks up more paper clips) when 1) the current in the wire increases with a stronger battery, and 2) the number of loops increases.

Conclusions:

_____

_____

_____

_____

_____

_____

_____

_____

_____

_____

_____

_____

_____

_____

_____

_____

Looking at the graph, help the students discuss their results and draw conclusions based on the data they have collected. Help them discuss any sources of error they might have encountered.

Review

Answer the following questions:

1. What makes materials magnetic?

an uneven number of spinning electrons

2. Why are some materials magnetic and others not? *only some materials*

have an uneven number of electrons spinning.

3. What are poles? *opposite ends of a magnetic*

4. Opposite poles (attract) or repel. [circle the correct word]

5. Like poles attract or (repel). [circle the correct word]

6. What happens if an electric current flows around a metal rod?
                    it induces a magnetic field

7. What happens if a magnetic rod is pushed and pulled through a wire coil?
                    it induces a voltage

Draw a magnetic field

NOTES:

Chapter 9: Light and Sound

Time Required:

Text reading - 30 minutes
Experimental - 1 hour

Experimental Pre-setup:

NONE

Additional Materials:

NONE

Overall Objectives:

In this chapter the students will be introduced to light and sound energy. They will look closely at waves and their properties, discuss the electromagnetic spectrum and visible light, and examine the difference between light waves and sound waves.

9.1 Light

Ask the students what they think light is. If a flashlight is available, have them explore some properties of light.

What happens to the light when you shine it in the dark? (*it shines in a straight line until it disappears or hits something*)

What happens to the light when you shine it during the daylight? (*it disappears immediately and you can't see it*)

What happens if you put your hand in front of the beam? (*the light is blocked*)

What happens if you shine it in a mirror? (*it gets reflected or "bounces back"*)

What happens if you shine it on a non-mirrored surface? (*it disappears into the surface*)

Can you "catch" the light in your hands? (*No*)

Can you catch the light in a bottle or a box? (*No*)

What happens to the light if you close your eyes? (*Nothing-we just can't see it*)

Can you hear light? (*No*)

From these observations they should understand that light cannot be "captured," that it bounces back (reflects in a mirror), that it "disappears" in sunlight because it blends into other light, that it can be blocked (by their hands), that it gets absorbed by non-reflective surfaces and that it cannot be detected by other senses -- such as touch or hearing.

Explain to the students that light is actually a combination of electric and magnetic fields. Light is an *electromagnetic wave*. Explain to them that because light is a wave -- it will have the properties of waves. (*Light also has particle-like properties, but this concept is outside the scope of Level I*).

9.2 Waves

Discuss with the students the nature of waves. Ask them to describe what they know about waves, what they look like, and how they move. If time permits, have them fill a bowl full of water. Add a small piece of styrofoam or something that will float. With an eye dropper, drop water droplets into the water. Have them observe how the water moves and how the small floating foam moves. They should see the water "ripple" and appear to bounce off the edges of the bowl and then go back to the center. The small piece of styrofoam will more or less stay in one place. Explain to the students that the water is mainly moving up and down and the disturbance from the water droplet moves outward.

Looking at the drawings in 9.2, describe to the students the parts of a wave. Show them that a wave has a "peak" and a "valley" and that the peaks are separated by a certain distance called a "wavelength." The height of a peak from the center is called the "amplitude."

Show them that the waves can be stretched or squeezed by moving the peaks farther apart or closer together. Explain that this changes the wavelength, but not the amplitude.

Explain to the students that each color of light is an electromagnetic wave with a different wavelength. They will look more closely at visible light in the next section.

Explain to the students that we can only see a small part of the electromagnetic spectrum. Radio waves and microwaves are light -- but we cannot see them.

Look carefully at the electromagnetic spectrum and show them that visible light is between the infrared light and ultraviolet light. Show them that radio waves are much longer than visible light and x-rays are shorter. Although we cannot see infrared light, we can often feel it. The heat from a campfire or from a glowing hot burner on a stove is mostly infrared light being absorbed by our skin and causing it to warm up. Likewise, we cannot see ultraviolet light, but it has effects we can detect. UV light from the sun is responsible for sunburns and suntans.

9.3 Visible light

Discuss with the students the different colors of light. For example:

What are the different colors in a rainbow?
(*red, orange, yellow, green, blue, violet*)

Are the colors ever in a different order? (*No*)

Do you ever see yellow, purple, red and then green in that order in a rainbow? (*No*)

When does a rainbow occur? (*when it rains*)

Do rainbows always occur when it rains? (*No*)

Explain to the students that each color of light is a wave at different wavelengths. Show them that red is the same as violet except that the wavelength for red is longer. Explain to the them that by squeezing a red wavelength or stretching a violet wavelength, they would become a different color.

Because colors have different wavelengths, when white light gets split the colors come out in a particular order. This is why the rainbow always goes from violet to red, red being the longest wavelength and violet being the shortest. A rainbow can be inverted (a double rainbow) but the colors will be in the same order.

White light

Violet short

Red long

9.4 Sound waves

Have the students think about sound. For example, ask them:

> What makes sounds? (*anything that crashes, birds, musical instruments etc.*)

> If you place your ear on a table and have a friend tap the other end, what do you hear? (*a vibration*)

> When you speak, what makes the sound? (*vibrating vocal cords inside your throat*)

> What does your throat do as you speak? (*it vibrates*)

Explain to the students that sound waves are not the same as light waves. Sound waves are waves of air particles. When they hear sound, the particles in the air or in a table are vibrating in a wave and we pick this vibration up with our ears to hear sounds.
Explain to the students that there are different sounds, just like there are different colors. A *frequency* is the number peaks that pass in a given time.

Low frequencies have long wavelengths and high frequencies have short wavelengths. We call the frequency of sound, pitch. High pitch means high frequency and short wavelength.

The intensity of a sound depends on its amplitude. Loud sound have high amplitudes and soft sounds have low amplitudes.

Discuss with the students the definition of a decibel. Explain to them that the human ear can be damaged by loud noises. This is why workers in factories or in airports wear earplugs around loud machinery or near airplanes.

9.5 Summary

Discuss the summary statements with the students.

Experiment 9 :   Bending Light and Circle Sounds   Date: _____

Objective: _____

              _____

Hypothesis   _____

Materials

       two prisms (glass or plastic)
       flashlight
       metal can open at both ends
       aluminum foil
       rubber band
       laser pointer
       long wooden craft stick

Experiment:
Part I: Bending Light

1. Take one prism and shine the flashlight through it at the 90° bend
   (see illustration). Record your results.

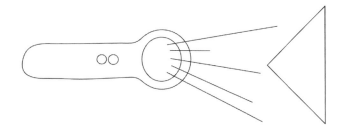

In this experiment the students will examine some properties of light and sound.

Have the students read the entire experiment and then write an Objective and a Hypothesis. Some examples are:

Objective:

    We will examine light through a prism to separate the colors.
    We will look at the wave nature of sound.

Hypothesis:

    Sunlight will not separate into different colors through a prism.
    Sunlight will separate into different colors through a prism.
    The flashlight will not separate into different colors through the prism.
    The flashlight will separate into different colors through the prism.
    We will see waves with the "soundscope."
    We will not see waves with the "soundscope."

Have the students shine the flashlight through the prism and record their results. Have them take the prism outside and shine sunlight through it. Have them note any differences. This can be tricky to do. Help them angle the prism so that the light will pass through it. The resulting rainbow will be cast ahead of the prism.

2. Now take the prism and shine sunlight through it from the same direction. Record your results.

3. Take the second prism and place it directly in front of the first one laying it flat on one of the short edges. Using the flashlight, shine light through the two prisms together (see illustration).

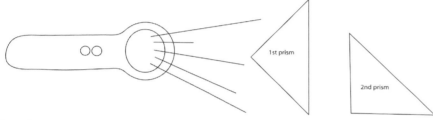

Results:
Part I.

1. What happens when you shine a flashlight through the prism?

_____

_____

2. What happens when you put the prism in sunlight?

_____

_____

Have them place two prisms together and shine the flashlight through both of them. Here they will observe a "double rainbow." The first rainbow will have red on the bottom and violet on top [or vice versa depending on how you shine the light through] and the second rainbow will be inverted -- that is, exactly opposite of the first rainbow. This can be hard to do, but help the students carefully adjust the angles and positions of the two prisms until the double rainbow is visible. Have them record their results.

Draw what you see

```
┌─────────────────────────────────────────────────────────┐
│                                                         │
│                                                         │
│                                                         │
│                                                         │
│                                                         │
│                                                         │
│                                                         │
│                                                         │
└─────────────────────────────────────────────────────────┘
```

Part II: Circle Sounds

1. Assemble a "sound scope" in the following way.
2. Take the metal can and make sure it is completely open on both ends.
3. Place a piece of aluminum foil over one end of the can and secure it with a rubber band. Be careful not to wrinkle the foil; try to keep it smooth.
4. Fix the craft stick to the metal can with strong tape.
5. Place the laser pointer with the light facing the foil on the craft stick. It should look like this:

Have the students assemble the soundscope. Make sure the aluminum foil does not become wrinkled.

When working with a laser pointer, have the students be very careful not to shine the beam in their eyes!

The laser beam should reflect off of the aluminum foil and onto an opposing wall. It will usually work best if the laser hits the center of the foil. Have them record what they see as they speak into the can. Have them make both high-pitched and low-pitched sounds.

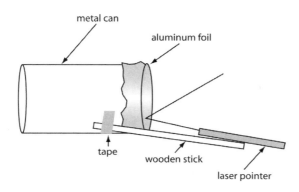

metal can
aluminum foil
tape
wooden stick
laser pointer

They should observe different-shaped circles. Explain to the students that these are waves -- but they look like circles because the laser pointer is stationary. If the laser pointer were moving these would look like normal waves. The waves or circles show the vibrations of the aluminum foil, which are caused by the sound vibrations in the air. This experiment allows us to visualize sound.

6. Turn on the laser pointer [Be careful not to point the laser directly into the eyes]. and observe the reflection on a wall or white board.
7. Holding the can to your mouth, speak into it and watch what happens to the reflected laser light. Record your results.
8. Continue to speak or sing into the can recording as many different shapes as you can.

Draw the shapes you see;

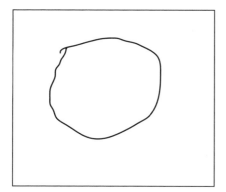

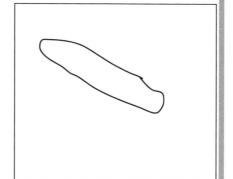

Review

Define the following terms:

electromagnetic wave *magnetic and electric fields combined*

wavelength *the distance between two peaks*

amplitude *the height of a wave from the middle point*

electromagnetic spectrum *radio waves to gamma rays*

visible light *the part of the electromagnetic spectrum visible to the human eye*

pitch *different wavelengths of sound*

frequency *the number of peaks that pass over a given time*

Are radio waves sound? *No- radio waves are light with long wavelengths*

NOTES:

Chapter 10: Conservation of Energy

Time Required:

> Text reading - 30 minutes
> Experimental - 1 hour

Experimental Pre-setup:

> NONE

Additional Materials:

> NONE

Overall Objectives:

In this chapter the students will connect the various types of energy they have been studying and learn about the Law of Conservation of Energy. This law states that energy is neither created nor destroyed but is just converted from one form to another. When we hear about an energy "crisis" it is referring to the depletion of usable energy, i.e. energy that can be converted into other forms of energy or work.

10.1 Introduction

Review with the students the types of energy they have studied and discuss with them how energy is converted from one form to another. For example:

graviational potential energy of the toy car converted into kinetic energy as it rolled down the ramp (Experiment 3).

chemical energy in a battery converted to light energy (Experiment 7)

chemical energy in a battery converted to magnetic energy and mechanical energy (Experiment 8)

electric energy converted into mechanical energy (Experiment 6)

10.2 Energy is conserved

Discuss with the students the drawing showing the toy car travelling down the ramp. Explain that when the toy car is sitting at the top of the ramp and is not moving, it has GPE but no KE. Show that, as the car rolls down the ramp, it picks up KE but loses GPE. Explain to them that it loses GPE because, as it rolls down the ramp, it loses height and as it loses height it loses GPE. Discuss with them the equation at the bottom of the graphic. Show them that, at each point on the ramp, the total energy (which equals GPE + KE) stays the same. Explain that this is what is meant by *conservation* of energy.

10.3 Usable energy

Strictly speaking, in the toy car example above, not all of the GPE is converted into KE. Because of friction, some amount of GPE is converted into heat instead. Explain to the students that heat is "unusable" energy. That is, the heat cannot be converted into another form of energy. We say that the energy is "lost," but we really mean it is not useful anymore. Have the students think about other examples where usable energy is "lost." For example:

A light bulb converts electrical energy into light (useful), but also gets hot (produces heat energy which is not useful). In the end the electrical energy and light energy are all "gone" -- converted into heat energy.

A car engine converts chemical energy in gasoline into useful mechanical energy, but eventually all of the mechanical energy gets dispersed as heat because of friction (from the air, engine, tires and brakes). In the end the gasoline is "gone" and the mechanical energy is "gone" too. All of the energy has been converted into heat.

A CD player converts electrical energy into music (useful), but also produces some heat. In the end the electrical energy is "gone" and can no longer play the music.

Eventually every kind of useful energy gets converted to heat. This is an example of the Second Law of Thermodynamics in action. The energy is not really gone but has been converted into a useless form.

Explain to the students that, when they hear the term "energy crisis," it means that usable energy (energy that can be converted into other forms of energy and do work) is being used up.

10.4 Energy sources

Discuss with the students different sources of energy. Explain to them that some sources of energy, like fossil fuels, cannot be renewed. That is, once they are used they cannot be replaced. Explain that fossil fuels come from plants and animals that died a long time ago. As the plant or animal decomposes, natural gas, oil, and coal get formed. Reservoirs of fossil fuels can be found in between the rock layers under the ground. The fuel can be mined by digging into the ground and removing the fossil fuel. Ask the students to list some fossil fuels and how they are mined.

For example:

coal -- mined from under the ground
*natural gas- found above oil and "mined" or brought to the
  surface with pipes*
oil -- pipes are drilled into the ground to remove the oil

Explain to the students that most of the energy we use comes from fossil fuels. Discuss why some day the fossil fuels will no longer be available. Ask them what we might do when the fossil fuels are gone.

Discuss with the students that there are renewable sources of energy such as solar energy, wind energy and energy that comes from water. Explain that some of these energy sources are expensive to use right now, but that it may be possible someday to use them with greater effecience instead of fossil fuels.

10.5 Summary

Discuss the summary statements with the students.

Experiment 10 :   On Your Own   Date: _____

You design this experiment. The goal is to convert as many forms of energy as you can into other forms of energy.

For example:

A senario can be designed so that energy is used to put out a fire. A marble is rolled down a ramp and bumps into a domino with a small cap of baking soda on top of it. A chemical reaction is started when the baking soda falls into the vinegar which produces carbod dioxide which then puts out the fire. In this case the rolling marble has kinetic energy which is used to convert gravitational potential energy into kinetic energy (the falling baking soda) which then starts a chemical reaction.

For this chapter the students will design their own experiment. The goal is to use as many different forms of energy as they can and to convert the different energies into other forms of energy.

In the example given, the kinetic energy of a rolling marble is used to knock down a domino that has a cap of baking soda on top. The baking soda falls into the vinegar and a chemical reaction is started.

In this example, the marble begins with GPE which gets converted to KE as it rolls down the ramp. The KE is used to convert the GPE of the elevated baking soda into KE as it falls. This releases the chemical potential energy (CPE) in the baking soda and vinegar and a chemical reaction starts producing CO_2 which puts out the fire. The chemical energy is converted into heat energy and bubbles.

Have the students think of ways this example might be extended. For example, the gas from the chemical reaction could be released into a small balloon or used to move a small piston.

First have the students do several "thought experiments" by asking themselves how they might set up a series of small scenarios, like the example.   Some of their ideas will not be practical, but have them use their imagination to think of ways to convert energy.

Using Energy to Put Out a Fire

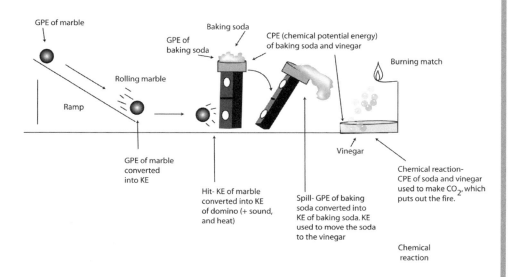

Use the following guide to design your experiment:

1. Write down as many different forms of energy you can think of.

kinetic energy _____   _____

_____   _____

_____   _____

_____   _____

2. Write down how these forms of energy can be represented.
kinetic energy _____

rolling marble   _moving toy car_   _moving ball_

_____

_____   _____   _____

_____

_____   _____   _____

_____

_____   _____   _____

Help them narrow some of their ideas into practical applications. Have them think of different forms of energy and how they might represent these different forms.

3. Write down ways to connect two or more of these forms of energy and explain how one form will be converted into another.

<u>moving toy car bumps into marble and starts it rolling</u>

Next, have them think about ways to link their different ideas. Have them start thinking about whether or not their ideas could work and have them start looking for items that they might use.

4. Now design an experiment to convert one form of energy into another. Give your experiment a title and write an objective and a hypothesis. Write down the materials you need and then set up a page to collect your results. See how many different forms of energy you can convert. Make careful observations and draw conclusions based on what you observe.

Experiment 10: _____ Date: _____

Objective: _____
_____

Hypothesis: _____
_____

Materials:

_____   _____   _____
_____   _____   _____
_____   _____   _____

Experiment:

_____
_____
_____
_____
_____
_____
_____
_____
_____
_____
_____
_____
_____
_____
_____
_____
_____

Have the students design their own experiment and write an objective and hypothesis. Help them assemble a materials list and then have them write out the steps of their experiment.

Results:

_____
_____
_____
_____
_____
_____
_____
_____
_____
_____
_____
_____
_____

Have them record their results whether or not their experiment worked. Have them write conclusions based on their results and ask them what they might do differently next time.

Conclusions:

_____
_____
_____
_____
_____
_____
_____
_____
_____
_____
_____

Extra page (use for graphing or recording more results)

This is an extra page for them to use for recording more results or for graphing any data they have collected.

Review

Describe the Law of Conservation of Energy

The law that states that total energy is conserved. This means that energy is neither created nor destroyed.

What is the energy that is conserved?

A. kinetic energy

B. potential energy

C. (total energy)

D. chemical energy

What is usable energy?

energy that can be converted into other forms of energy or used to do work.

Name one form of energy that is sometimes unusable.

heat energy or light energy

NOTES: